SPOTLIGHT

THE AGE OF EXPLORATION AND DISCOVERY

Leonard W. Cowie

Wayland

SPOTLIGHT ON HISTORY

Spotlight on the Age of Exploration and Discovery
Spotlight on the Age of Revolution
Spotlight on Elizabethan England
Spotlight on the First World War
Spotlight on the Industrial Revolution
Spotlight on Post-War Europe
Spotlight on the Second World War
Spotlight on the Victorians

First published in 1985 by
Wayland (Publishers) Ltd
61 Western Road, Hove
East Sussex BN3 1JD

British Library Cataloguing in Publication Data
Cowie, Leonard W.
Spotlight on the age of exploration and discovery—(Spotlight on history)
1. Voyages and travels—Juvenile literature
I. Title
910′.9′03   G175

ISBN 0–85078–614–2

Typeset, printed and bound
in Great Britain at
The Bath Press, Avon

CONTENTS

1 The Eve of Exploration   4

2 Prince Henry's Mariners   11

3 The Opening Up of the East   18

4 The Discovery of the New World   25

5 The Conquistadores   32

6 The Challenge of the Dutch   40

7 The English in North America   47

8 The French in Canada   56

9 Overseas Rivalry   65

Glossary   71

Date Chart   72

Further Reading   74

Index   75

1 THE EVE OF EXPLORATION

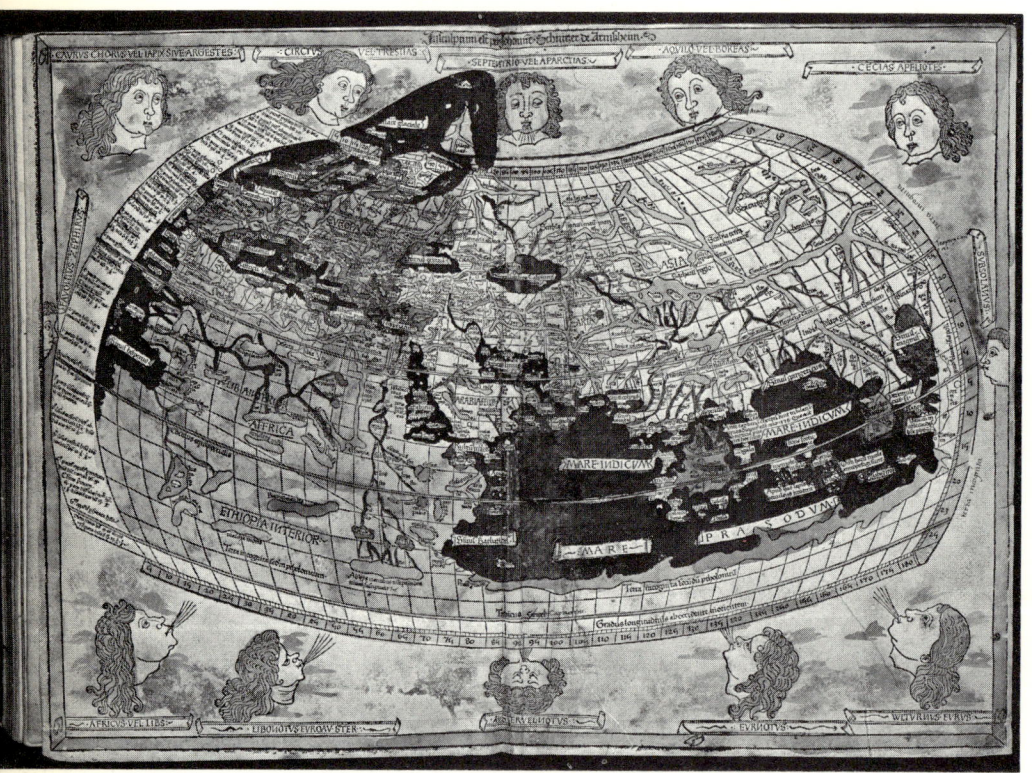

This Ptolemaic map shows how the world was viewed prior to the age of exploration.

Within a period of less than three hundred years, from about 1420 to 1713, European explorers discovered that all the seas of the world formed one continuous ocean, a fact which encouraged them to undertake voyages into previously uncharted areas. They found new lands and opened up new trade routes. Colonies and settlements were established in many parts of the world. Strange products and great wealth were brought back to Europe. Nations became powerful through their acquisition of overseas territories and consequently fought each other for supremacy over them.

Sailors feared the monsters which they imagined inhabited unknown waters.

The fearsome sea

In the centuries before the age of discovery and exploration, the most important countries of Europe were those bordering the Mediterranean sea, through which lay the main trading routes linking east to west. Ships did not venture far in other waters, as little was known of them and sailors were afraid of the many dangers they might meet.

The Mediterranean is a comparatively sheltered and calm sea, and, it was generally expected that if a ship sailed out into the open ocean far from the protection of land, it would meet terrible, violent storms. These storms were believed to be so strong that they could sink ships in an instant, tear apart driftwood, and drown men by dragging them down into the waters. In addition, these regions were rumoured to be prone to fogs, which were only a rare occurrence in the Mediterranean. Fogs presented dangers both real and imaginary: they made it impossible for ships to steer by the sun and stars, and were also suspected of carrying infectious diseases, which would spread among those who were surrounded by them. Shakespeare referred to this belief when he described in *A Midsummer Night's Dream* how 'the winds . . . have sucked up from the sea contagious fogs.'

Those were not the only fears. Sailors believed that if they braved these storms and fogs, they would go into regions where they would meet with certain death. If they journeyed southwards down the Atlantic coast of Africa, beyond Morocco, they would come close to the Equator. Here, according to learned scholars, the sun's rays came down in liquid flames, which made the sea boil, set ships alight, and turned men into blackened corpses. If they travelled westwards out into the Atlantic, they might sail into the 'green sea of darkness'. According

According to learned scholars, scenes such as this lay in wait for sailors who travelled southwards towards the equator.

to Arab geographers, this was a watery swamp where ships would become stuck and fall prey to the terrible, fierce monsters lurking there, which were capable of overturning a ship with ease. If they sailed northwards, they would enter a frozen waste where they would come upon Judas, the betrayer of Christ, who lingered near the mouth of hell, too wicked to be admitted even there.

The urge to explore

Why was it that in the fifteenth century the Europeans overcame these powerful fears and were ready to risk the awesome perils described in these legends? One of the Spanish explorers said that they set out on their journeys of discovery, 'to serve God and His Majesty, to give light to those who were in darkness and to grow rich, as all men desire to do.' And those were the aims of them all—to gain wealth and to spread Christianity.

Wealth could be gained by trade. The richest trade in Europe was in silk, ivory, precious stones, gold, silver and, above all, spices (such as pepper, cinnamon, ginger and cloves), which Europeans needed to make palatable the salted meat upon which they lived during the winter. These items came mostly from the East by way of the long overland route, but the trade was controlled by the merchants of the Italian republics of Venice and Genoa, who bought the goods from the Arabs at Alexandria. Other Mediterranean countries envied the prosperity that this trade produced, and were anxious to see if they could open up routes that would enable them to share it.

The explorers were also religious men. The Venetian traveller, Marco Polo, had made his way overland to the empire of the great Eastern ruler, Kublai Khan, in the thirteenth century. He had brought back accounts of the peoples of India, China and Japan, who were civilized and wealthy but practised pagan religions. Beyond the seas there were likely to be many lands inhabited by heathens whose souls must be saved. Christopher Columbus had read Marco Polo's account of the east. He wrote to the Spanish court about 'what I conceive to be the principal wish of our serene King, the conversion of these people to the holy faith of Christ.'

The travels of Marco Polo 1271~95

Navigation, shipbuilding and gunnery

These voyages of exploration, however, would not have been possible without the aid of a number of new inventions and discoveries that were made in Europe at this time. The most significant of these were in three important branches of knowledge.

One of these was the study of geography and astronomy, especially as applied to the problems of sailing-ships. The fifteenth century saw the emergence of the Renaissance in Europe, a time when the writings of antiquity, which had been lost since the fall of the Roman Empire, were rediscovered. These included the works of geographers, who calculated that the world was a sphere and consisted of continents surrounded by water. There were also Greek almanacs, which showed how to predict exactly the positions of the sun, moon, planets and all the fixed stars used in navigation. The European explorers needed also to be able to measure accurately the height of heavenly bodies to gauge their position precisely; one sailor, for instance, could only

Measuring the distance from ship to shore, using a quadrant.

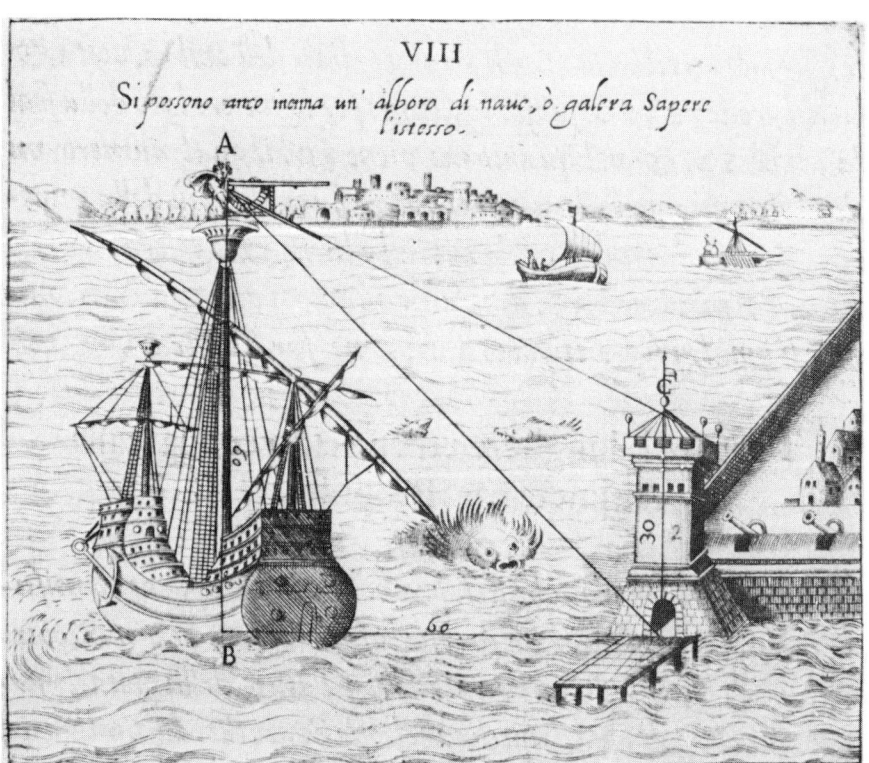

say that the Pole Star appeared to be 'about a third of a lance above the horizon.' Mariners could now make more accurate measurements by using a newly invented instrument called the quadrant, which when used in conjunction with the compass (invented in about 1250), allowed an exact course to be steered.

There were also important changes in shipbuilding methods. European ships in the Middle Ages had a single mast carrying a square

A Portuguese 'caravel' with a lateen sail.

The 'carrack' was the original galleon. The square rig of its sails was far more suitable for running before the ocean trade winds than the lateen of the 'caravel'.

sail and could not sail into the wind. The Arabs overcame this restriction by devising the lateen, a small triangular sail set at an angle to the mast. During the fifteenth century Portuguese and Spanish ship-designers produced first the caravel, and then the carrack, which had both square sails and the lateen. Ships were also built longer than before and so were more seaworthy.

Finally, it was essential that ships making voyages into unknown regions were armed. In the fifteenth century the Venetians mounted small guns along the decks of their ships, and early in the next century the Portuguese designed war vessels with a series of openings on each side, through which big guns could be fired. It was the Portuguese who first sank ships by using gunfire, when they defeated the Arabs in Indian waters.

2 PRINCE HENRY'S MARINERS

Portugal is a narrow strip of territory that lies on the Atlantic coast, to the west of Spain, and is a little larger in size than Ireland. In the fifteenth century, because it was cut off from the trade routes of the Mediterranean, it had to support itself by turning to the open sea. It established an important shipbuilding industry and engaged in fishing and whaling in the Atlantic.

A statue of Prince Henry The Navigator, 'The Father of Atlantic Exploration'.

Prince Henry the Navigator (1394–1460)

These activities made it possible for Portugal to begin a great movement of overseas exploration, setting an example which was followed by the other major sea-powers of Europe. The leader of this enterprise was Prince Henry 'the Navigator', the son of the King of Portugal,

Prester John, a legendary Christian priest and king, believed in the Middle Ages to have ruled in the Far East.

and whose mother was an English princess. A historian of the time said of him, 'Strength of heart and keenness of mind were in him to a very excellent degree, and beyond comparison he was ambitious of achieving great and lofty deeds.'

It was during the Portuguese capture of Ceuta, on the northern Moroccan coast, in 1415 that Prince Henry's ambitions first took shape. In the bazaars and storehouses of the captured town he saw valuable wares that had been brought from the East by the overland route across the desert. He also heard of the legendary kingdom beyond the desert, which was said to have a Christian ruler, Prester John. Portugal was too small a country to conquer more territory in Africa, but Henry was sure that she could use her sea-power to explore the African coast and discover the starting point of this desert route, which he believed was near the Gulf of Guinea. If he could achieve this he could fulfil two objectives: he could win for Portugal a share in the trade in gold, ivory, pepper and slaves, and he could also form a partnership with Prester John to launch a joint crusade against the Muslims in North Africa and the Holy Land.

The 'Prince's Town'
The idea of discovery and conquest was of such importance to Henry that he resolved to devote his whole life to it. In 1419, he left the royal palace in Lisbon and settled in an old fort by the rocky shores of Sagres Bay, on the southern tip of Portugal. To this fort he added a small chapel, a study and an observatory, and soon the site became known as the 'Prince's Town'.

His aim was to establish there a sort of seafaring academy, which would prepare Portuguese sailors as fully as possible for the voyages on which he intended to send them. A Portuguese historian said, 'In his wish to gain a prosperous result for his efforts, the Prince devoted great industry and thought to the matter and at great expense procured the aid of one Master Jacome from Majorca, a man skilled in the art of navigation and in the making of maps, and who was sent for, with certain of the Arab and Jewish mathematicians, to instruct the Portuguese in that science.' Indeed, Henry brought a range of experts from many countries to his academy, where they could continue their studies and pass on their knowledge to his mariners.

Henry also wanted to ensure that only the most up-to-date ships were used on expeditions. He established shipyards at the port of Lagos, near to Sagres, where caravels were specially designed and built for making long ocean voyages. An Italian seaman said that the 'caravels of Portugal were the best sailing-ships afloat'. So Henry provided his sailors with the knowledge, instruments and ships to ensure their success.

Prince Henry The Navigator, planning his voyages from his castle near Cape St Vincent in Portugal.

The voyages of the caravels

The first of Prince Henry's expeditionary forces sailed down the African coast in 1420. It was caught in a storm and driven to a small island which was named Porto Santo ('Holy Haven') by the crew, 'at this very time in their joy at thus escaping the perils of the tempest.' This proved to be one of a small group of islands called Madeira, which Henry set about colonizing. New crops were introduced there, such as sugar-cane from Sicily and grapes from Crete, and even rabbits were brought from Portugal.

In 1428, after several fruitless voyages, Henry's elder brother returned from Venice with a map that had been made in 1351 and which marked the position of the Azores. In 1431, Henry sent out an expedition to these islands which were situated a thousand miles out into the Atlantic. They too were colonized. Both Madeira and the Azores provided the Portuguese with useful bases for supplying and sheltering ships, and their temperate climate was healthier for Europeans than the African mainland.

The 'caravel', in which the Portuguese made their first journeys along the coast of Africa.

15

So far the Prince had found no difficulty in persuading his mariners to make their voyages. They were confident in the ships and equipment he had given them. 'Our sailors', said one of them, 'went out well taught and provided with instruments and rules which all map-makers should know.' Now, however, as he sent them further and further down the African coast, fears aroused by the old legends were re-awakened. One ship, sent out in 1432, came upon Cape Bajador, a long reef situated less than a thousand miles from Sagres and which jutted out to sea from the Sahara desert. When the sailors saw the water swirling around the Cape, they were convinced that the sea was boiling, just as the legends of the tropics had foretold. They insisted that the captain turn the ship back to Portugal.

In 1434 Henry sent the same captain to this region once again, but this time with a different crew. The ship rounded the Cape, and the captain told Henry that the sea there was 'as easy to sail in as the waters at home.' In the following year another ship sailed 390 miles beyond the Cape. In 1441 the Cape Verde Islands were discovered and occupied, and four years later Senegal was reached. By the time Henry died, in 1460, his caravels had sailed as far as Sierra Leone. This was only a third of the way down the coast of Africa, but the movement he had begun was to continue in the following years.

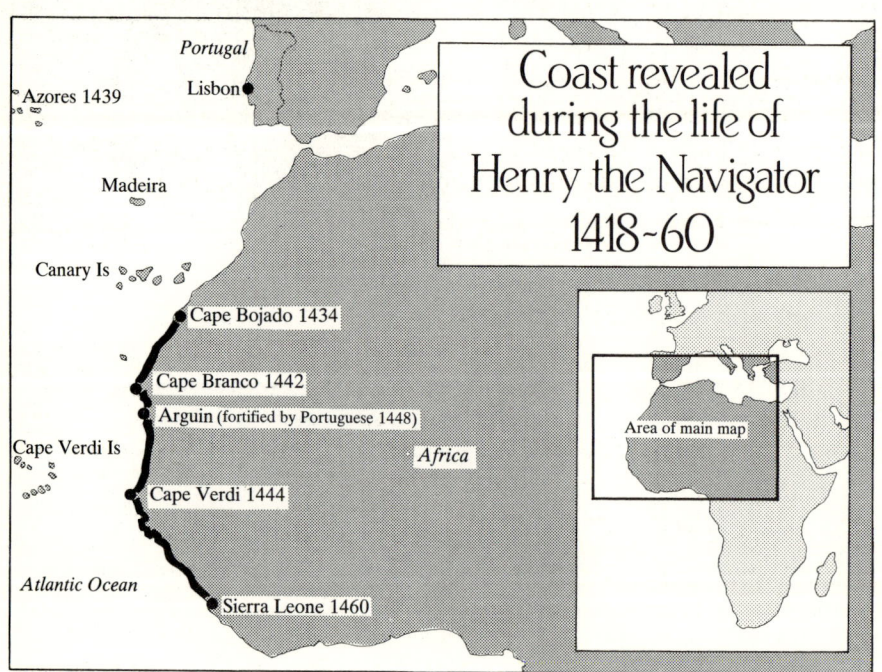

Portugal
Azores 1439
Lisbon
Madeira
Canary Is

Coast revealed
during the life of
Henry the Navigator
1418~60

Cape Bojado 1434
Cape Branco 1442
Arguin (fortified by Portuguese 1448)
Cape Verdi Is
Africa
Cape Verdi 1444
Area of main map
Atlantic Ocean
Sierra Leone 1460

Ships brought back cargoes of slaves from Africa. The notorious African slave trade was to continue for four centuries.

Prince Henry's ships had done more than explore. During the last twenty years of his life, they had stopped along the coast and sailed into rivermouths to trade with local chieftains. This trade gradually led the Portuguese nation to share Henry's interest in African exploration. The ships were bringing back the cargoes that the Prince had originally sought, including slaves, and in 1454 Pope Nicholas V authorized the Portuguese 'to attack, subject and reduce to perpetual slavery the Saracens, pagans and other enemies of Christ southward from Cape Bajador . . ., including all the coast of Guinea.' The notorious African slave trade, which was to continue for nearly four centuries, had begun.

17

3 THE OPENING UP OF THE EAST

After Prince Henry's death, Portuguese exploration of the African coast slowed down, though in 1473 the Equator was crossed without any sailor bursting into flames. Not until the reign of King John II (1481–95) was there any great revival in exploration, and by then the Portuguese had developed a flourishing trade with West Africa.

Bartholomew Diaz and Vasco da Gama
The climax of John's reign came in 1487, when Bartholomew Diaz set out with two ships and a supply vessel stocked with extra provisions for a long voyage. He was overtaken by a storm which blew him away from the sheltered waters near land and was driven southwards for thirteen days. When the wind finally abated, he sailed eastwards, in

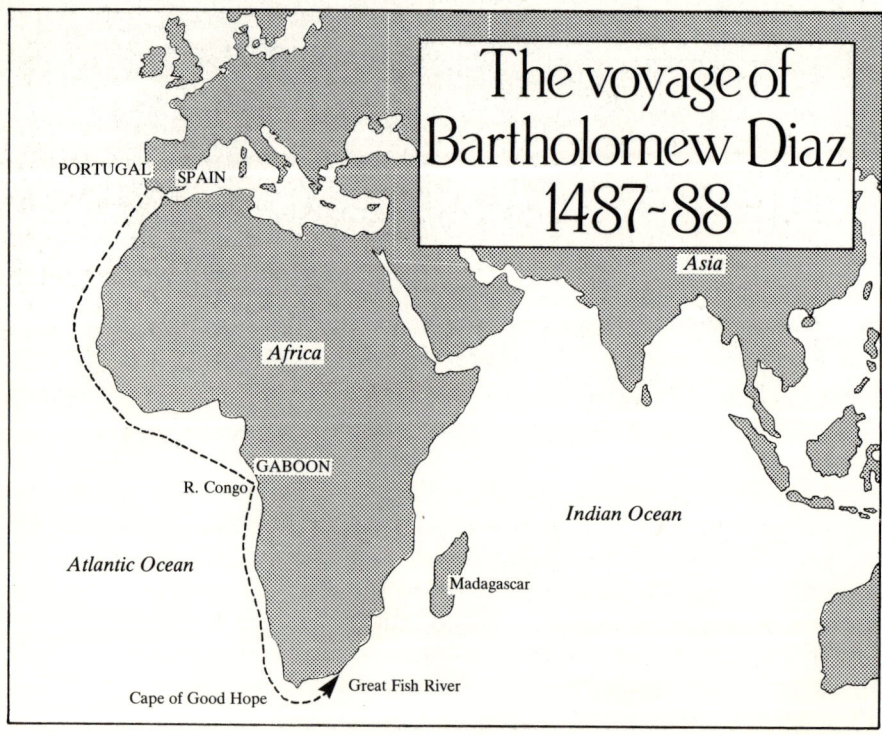

The voyage of Bartholomew Diaz 1487~88

PORTUGAL SPAIN

Asia

Africa

GABOON

R. Congo

Atlantic Ocean

Indian Ocean

Madagascar

Cape of Good Hope   Great Fish River

Admestor, the Phantom of the Cape of Good Hope, appearing to sailors.

the hope of finding land again, but when this proved unsuccessful, he turned to the north and saw mountains on the horizon. He found that the coastline turned away north-eastwards. Diaz was convinced that he had reached the further side of Africa and that beyond lay India and the East, but his men mutinied and forced him to turn back. On his return voyage, he discovered the southern tip of Africa, which he called the Cape of Storms, but King John later renamed it the Cape of Good Hope.

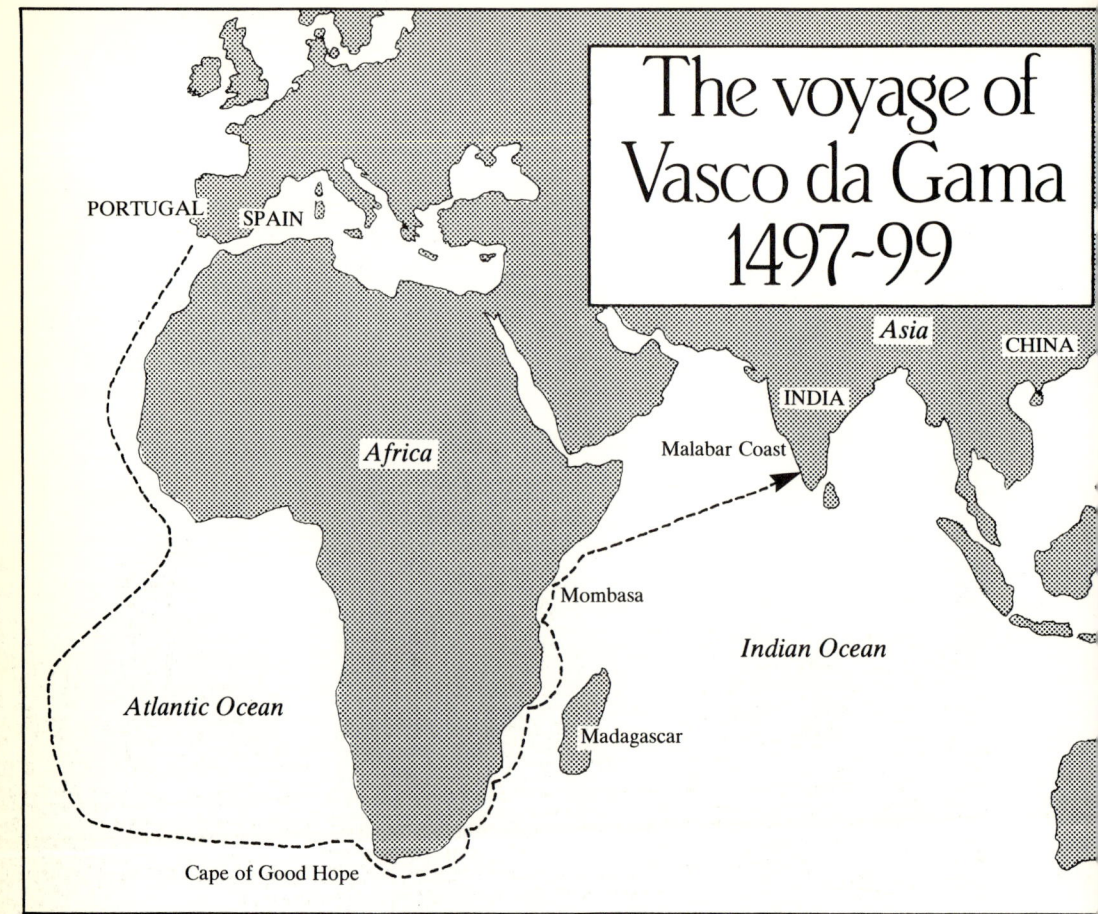

The voyage of Vasco da Gama 1497-99

PORTUGAL
SPAIN
Asia
CHINA
INDIA
Africa
Malabar Coast
Mombasa
Indian Ocean
Atlantic Ocean
Madagascar
Cape of Good Hope

In that same year, a Portuguese expedition followed the course of the River Nile through Egypt and reached Ethiopia, where they learned of Arab ships trading along the East African coast. Portuguese merchants were keen to sail in these waters and go on to India. In 1492, however, Christopher Columbus crossed the Atlantic and was thought to have discovered a western sea-route to India for Spain. The Portuguese felt that they would not have sufficient sea room for their voyages. A fleet was nearly sent across the Atlantic to challenge the Spanish claim, but conflict was averted by Pope John II. He persuaded the two countries to accept the Treaty of Tordesillas, which drew a line of demarcation down the Atlantic depriving the Portuguese of access to the West Indies, but reserving for them the exclusive rights to Brazil.

By 1495, when Manuel I became King of Portugal, it had become apparent that Spain did not possess a western route to India, and the Portuguese decided to renew the search for the eastern route. In 1497 Vasco da Gama set out from Lisbon equipped with 'four vessels to make discoveries and go in search of spices.' He sailed round Africa and across the Indian Ocean to reach Calicut after a voyage of ten months. The quest for the sea-route to India had been accomplished. When asked by the Indians what they wanted, the Portuguese replied, 'Christians and spices'. Calicut was a leading commercial city, and da Gama loaded his ships with spices in exchange for the merchandise he had brought, which was mainly good-quality iron. He returned home, having lost two ships and two-thirds of his men, mostly from scurvy, but on his cargoes 'the return was sixtyfold'.

The Samorin of Calicut gives an audience to Vasco da Gama.

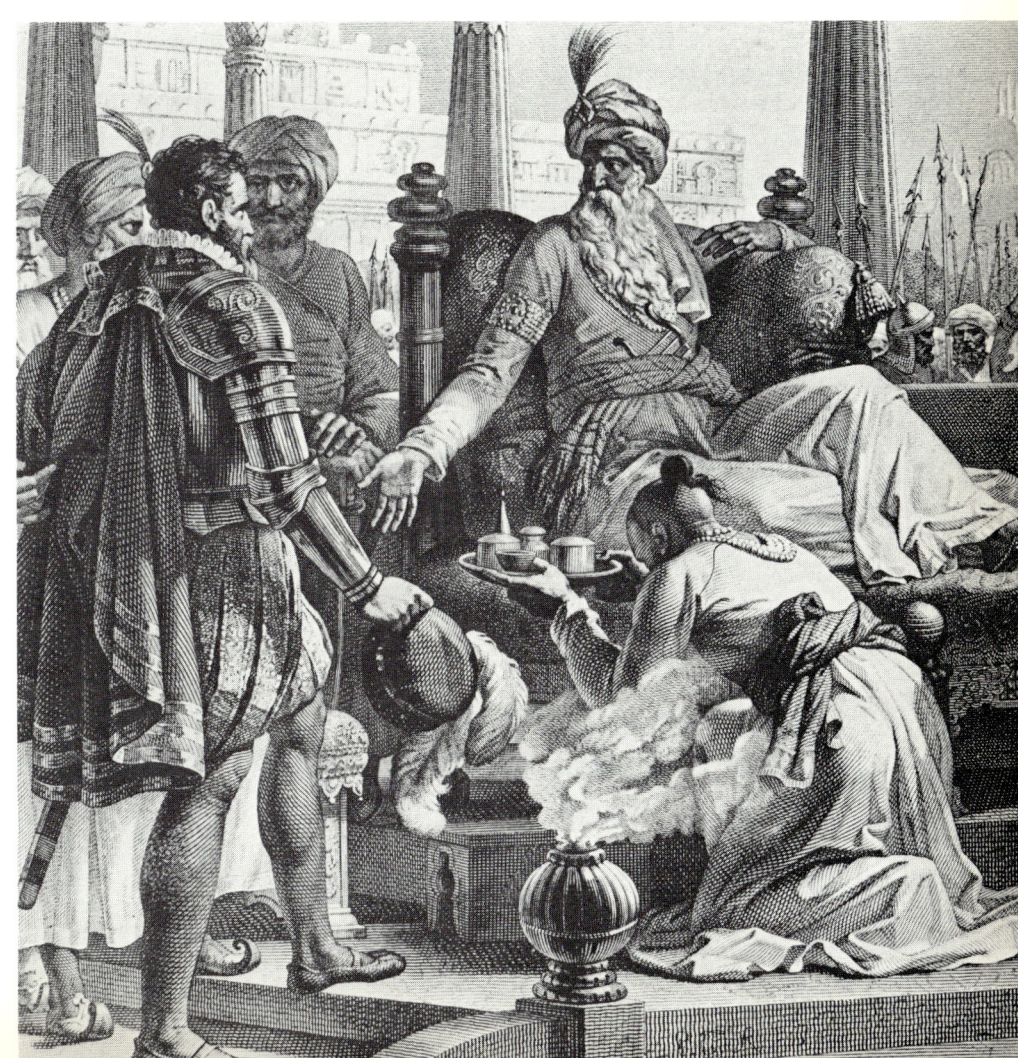

Trade and power in India

The opening of trade routes to India by the Portuguese filled Manuel with great enthusiasm. He wrote exultantly, 'As the principal motive of this enterprise has been ... the service of our God and Lord and our own advantage, it pleased Him in His mercy to speed them on their route. Henceforward all Christendom, in this part of Europe, shall be able ... to provide itself with these spices and precious stones.' In 1500 he gave himself the title of 'Lord of the Conquest, Navigation and Commerce of India, Ethiopia, Arabia and Persia.'

In that same year Manuel sent out a fleet of thirteen ships which sailed across the Atlantic under the leadership of Pedro Cabral, to claim Portuguese possession of Brazil. On their arrival one of the sailors wrote, 'There is great plenty, an infinitude of waters. The country is so well favoured that if it were rightly cultivated it would yield everything because of its waters.' But Cabral was not interested in settlement in South America; he wanted to develop the profitable Indian trade. He sailed back across the South Atlantic, rounded the Cape and arrived off Calicut, where his ships were loaded with a large cargo of pepper and other spices.

Regular trading-fleets were now despatched to India. Within three years the price of pepper at Lisbon was only a fifth of what it was at Venice. The Indians were prepared to provide constant supplies in return for cloth, gold, silver and copper. The Arab merchants resented the way in which the Portuguese were depriving them of their accustomed trade, and they began to harrass them. In 1505 Manuel

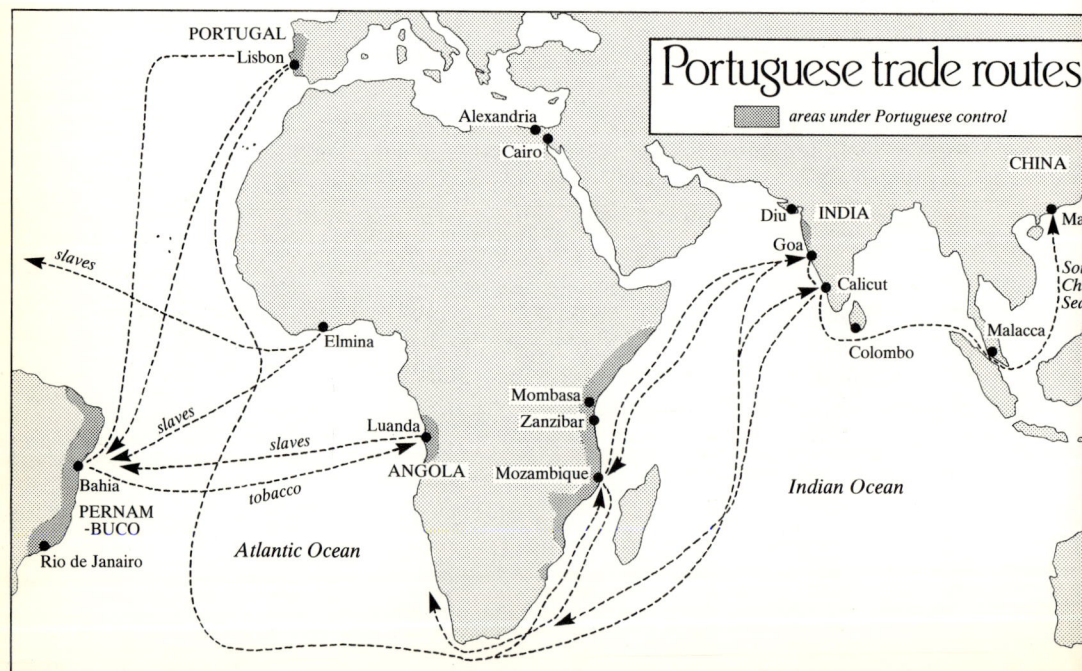

appointed Francisco del Almeida as his Viceroy in the East to uphold Portuguese power. 'Let it be known,' said Alemeida, 'that if you are strong in ships, the Indies are yours' and in 1509, the heavily-armed Portuguese ships destroyed their Arab rivals. Almeida's successor, Alfonso d'Albuquerque, saw the need for fortified shore-bases for the Portuguese fleet, but he insisted, 'My will and determination is, as long as I am Governor of India, neither to fight nor to hazard men on land except in those parts wherein I must build a fortress.' The Portuguese wished to come to India as traders and not as conquerors or settlers.

The Portuguese empire in Asia

Albuquerque consolidated Portuguese trading power; he conquered Goa, which he made the seat of Portuguese government, he gained control of the Straits of Malacca, which connected the Indian Ocean with the South China Sea; and captured Ormuz, the key port for trading in the Persian Gulf. His successors established trade with Japan in 1542, and with China in 1557. The Portuguese monarchy claimed a monopoly over all trade.

Trade with the East was organized on a regular, yearly basis. Each March, a fleet left Lisbon carrying not only the goods meant for trading purposes, but also the arms and supplies for the fortresses and trading-posts. This fleet arrived at Goa in September and left either in December, if it were to sail along the East African coast, or in January if it sailed directly to the Cape. When it arrived back in Lisbon, its cargoes were sent to Antwerp for sale throughout Europe, mainly by German traders.

The Portuguese also established churches and Mission stations in their empire. Their greatest evangelist was Francis Xavier. He belonged to the Society of Jesus, or Jesuits, whose General told him, 'God has given you the Indies—a whole world of people and nations. Kindle those unknown nations with the flame that burns within you.' Xavier arrived at Goa in 1542, and for ten years laboured among the peoples of the East Indies, as well as founding a mission in Japan.

By the middle of the sixteenth century, however, the empire was having disastrous effects upon Portugal. Her population was small, and too many people were attracted by the hope of gaining easy riches in trade. Domestic farming declined and depended on a work-force of African slaves. It was estimated in 1551 that a tenth of Lisbon's population of 100,000 were slaves, and that there were over sixty slave-markets. Moreover, their increased wealth brought about dishonesty and incompetence among the Portuguese, and consequently their power declined. The arrival of the Dutch in the East, in 1595, led to the destruction of their empire.

Francis Xavier the 'Apostle of the Indies' was sent by John III of Portugal as a missionary to the Portuguese colonies in the East.

4 THE DISCOVERY OF THE NEW WORLD

During the years when Portugal was conducting exploratory voyages down the African coast, her most powerful neighbour, Spain, was warring against the Moorish kingdom of Granada, which it did not defeat until 1492. By this time Portugal had established important trade links with West Africa and was likely to discover a sea-route to India. Spain resented her neighbour's achievements overseas and wished to rival them herself.

Christopher Columbus, the Italian navigator and explorer in the service of Spain, who discovered the New World in 1492.

The 'Enterprise of the Indies'

Several fifteenth-century cartographers believed that, if the world were round, it would be possible to sail west and eventually reach the eastern lands which Marco Polo had visited. Christopher Columbus, a Genoese seaman, was convinced that he could cross the Atlantic and confirm this belief. He tried to secure support for this 'Enterprise of the Indies' from several European rulers, but not until 1492 did he find King Ferdinand and Queen Isabella of Spain ready to supply him with ships and men for such a voyage. He was granted the title of Admiral and undertook 'to discover and acquire islands and mainland in the Ocean Sea.'

He sailed on 3 August 1492 with three carracks, the *Nina*, the *Pinta* and the *Santa Maria*, which was his flagship and the largest, though it was not much bigger than most modern trawlers. After taking in supplies at the Canary Islands, which Spain had occupied in 1479, the fleet sailed westwards on 6 September. When all sight of land disappeared, 'the sailors loudly lamented their fate and cried and sobbed like children'.

The winds were favourable, and the ships sailed on smoothly, day after day, but Columbus had thought that India would lie not far beyond the Canary Islands. Now food became short, and the crews were afraid. When they encountered the huge areas of floating weed in the Sargasso Sea, they believed it to be the 'green sea of darkness'. On 10 October they compelled Columbus to promise to turn back if no land appeared in three days.

Columbus' three ships the 'Maria', the 'Santa Maria' and the 'Pinta'.

Columbus and his crew land on San Salvador (now called Watlings Island) and are greeted by the native inhabitants.

On the morning of the next day, however, they saw some floating twigs and a piece of carved wood, and that evening a moving light was spotted in the darkness. At dawn they sighted a stretch of land which 'was rather large and very flat ... and the whole so green that it is a pleasure to look upon'. Columbus rowed shorewards bearing the Spanish royal standard and a large wooden cross, and 'as soon as the Admiral came to the shore, he fell on his knees and, weeping abundantly tears of joy, he began to say, "We praise Thee, O God, We acknowledge Thee to be the Lord."'

Columbus called the land San Salvador, which now forms part of the Bahamas and is named Watling's Island. He wrote in his journal, 'I intend to go and see if I can find the island of Japan'. Having sailed so far, he was sure it could not be far away. For the next three months

The wreck of the 'Santa Maria' on the coast of Hispaniola, in December 1492

he sailed from island to island. When he found Haiti and Cuba, he hoped that they were Japan, but there were no cities or silk-clad sages as Marco Polo had described. The light-brown people wore little clothing and live in thatched huts. When the *Santa Maria* finally ran aground, Columbus returned home with his two other ships, believing he had found islands off the Indies.

The failure of the western route

Columbus arrived back in Spain on 15 March 1493 and received a royal welcome. Everyone there believed that he had found the islands in the East that had been described by Marco Polo as being inhabited by savages who 'have no king or chief, but live like beasts.' He was allowed to make three more voyages (1493–96, 1498–1500 and 1502–04), during which he discovered Jamaica, Trinidad, the South American mainland and the Gulf of Mexico. He was still trying to find a way through to Asia, which he continued to believe lay only a little further to the west. On his last voyage, when he coasted the Isthmus of Panama, he declared he was 'within ten days sail of the Ganges River'. Until his death in 1506, he was still certain that he had almost completed the western route to the East, but his failure led those who had previously supported his efforts to desert him. 'If I had stolen the Indies and given them to the Moors,' he said bitterly, 'Spain could not have shown me greater enmity.' Despite his humilia-

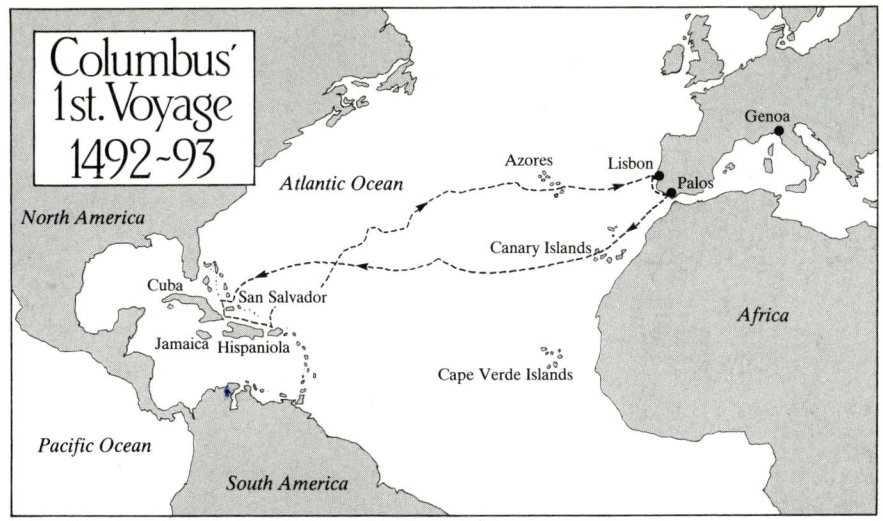

Columbus'
1st. Voyage
1492~93

Genoa
Azores
Lisbon
Palos
Atlantic Ocean
North America
Canary Islands
Cuba
San Salvador
Jamaica Hispaniola
Cape Verde Islands
Pacific Ocean
Africa
South America

tion and frustration, Columbus wrote memoir upon memoir on the Indies and their gold.

Columbus was, of course, quite mistaken in his beliefs. The Indies did, in fact, lie to the west of the lands he discovered, but they were 10,000 miles away beyond a great continent and an immense, uncharted ocean. Indeed, some years before his death, there was a growing belief that the new lands on the other side of the Atlantic were very much further from the mainland of Asia than he had estimated. As early as 1494, a cartographer, Peter Martyr de Angleria, on hearing reports of Columbus's first voyage, had stated that, 'When treating of this country, one must speak of a new world, so distant is it and so devoid of civilization and religion.'

The American continent
After the death of Columbus, more and more seamen sailed across the Atlantic, and they became certain that the land they found was not Asia. Among them was an Italian, Amerigo Vespucci, who made four voyages in the service of Spain and Portugal between 1497 and 1504. He said that he had seen a continent which 'it is proper to call a new world'. This inspired a German cartographer in 1507, to label the partly-discovered new continent as 'America', and this name was adopted for both the northern and southern part of the land-mass. The islands to which Columbus had sailed became known as the West Indies.

The fading belief that Asia might yet be only a short distance beyond the new continent was finally removed by a Spanish explorer, Vasco

Vasco Nunez de Balboa sets his dogs on the native inhabitants of Panama, claiming the territory for Spain.

Nunez de Balboa. In 1513 he penetrated the dense jungle of the Isthmus of Panama, where he heard from the Indians of a great salt sea lying a few miles to the south. Climbing a mountain peak in Darien he saw the 'Great South Sea'. Four days later—wearing full armour, brandishing a sword in one hand and the banner of Spain in the other, he waded into its waters to claim the sea and all adjacent lands for his country. It was now clear that Asia lay beyond yet another ocean.

Nevertheless, there was still the possibility that Asia could be reached by sailing across this ocean. In 1519 the Spanish government accepted the offer of a Portuguese explorer, Ferdinand Magellan, to seek a way round the southern part of the American continent to the East Indies. Magellan's voyage, which cost him his life in the Philippines, did not, however, bring Spain the desired prize. It took him over a month to sail the four treacherous miles of the stormy straits separating South America from the islands of Tierra del Fuego. 'Even if we have to eat the leather wrappings from the masts and yards,' he declared, 'I will go on.' When at last his ship reached the calm waters of the South Sea, he renamed it the Pacific Ocean, but it was clear that the Straits could not be used as a regular channel for trading ships. Instead of a sea-route to the East, Spain had gained a vast new continent to colonize.

Magellan discovers the stormy straights which bear his name.

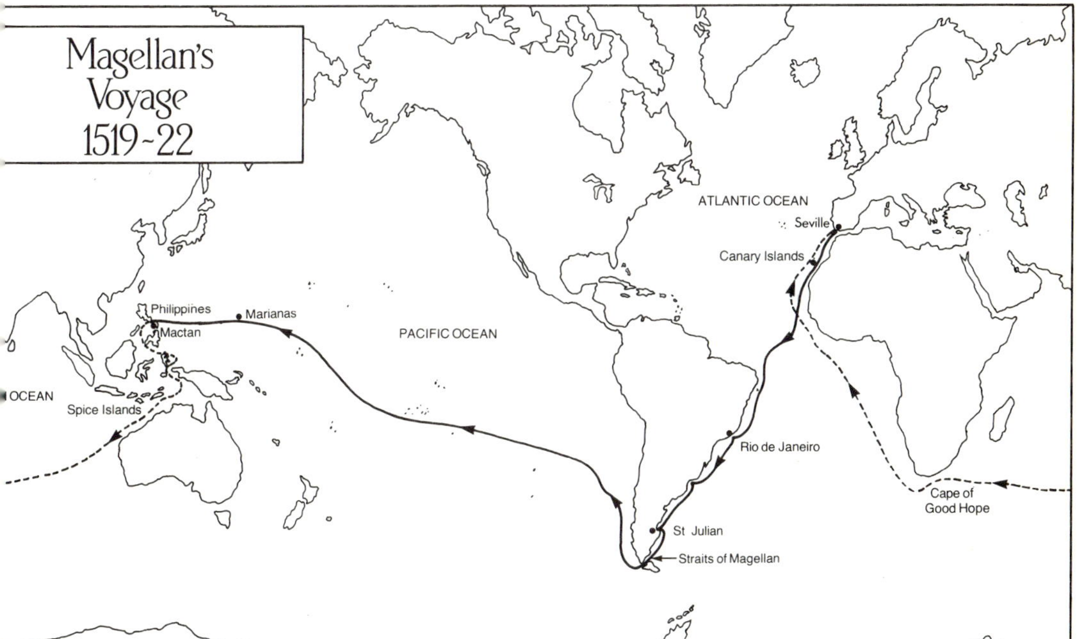

Magellan's
Voyage
1519~22

ATLANTIC OCEAN

Seville

Canary Islands

PACIFIC OCEAN

Philippines   Marianas

Mactan

OCEAN

Spice Islands

Rio de Janeiro

Cape of
Good Hope

St Julian

Straits of Magellan

5 THE CONQUISTADORES

When Ferdinand and Isabella agreed in 1493 that Columbus should make a second voyage across the Atlantic, they furnished him with a fleet of seventeen ships, carrying neither guns, nor trade goods, but rather 1,200 people equipped with tools, seeds and animals to be settled in Haiti. Columbus accepted Portuguese and Italian notions that supposed that any discoveries of other lands would be followed by peaceful trade. However, the Spaniards, inspired by their successful crusading campaigns against the Moors, believed in military conquest, division of lands, and the conversion of infidels.

The Spanish Conquistadores were professional soldiers who established the Spanish Empire in South America by ruthlessly conquering the Aztecs and the Incas.

Hernando Cortes leads the Conquistadores into Mexico.

By the middle of the sixteenth century, the Spanish Empire in South America had been established. This was achieved through the conquest of the two native empires of the Aztecs and the Incas by groups of adventurers known as the 'conquistadores'. They belonged to the lesser Spanish nobility and sought their fortunes in the New World. Their only skill was in fighting, and they had taken part in the war that had recently been concluded against the Moors. They were tough, professional soldiers, ruthless and brutal, but their campaign was also a 'spiritual conquest', which sought to replace the native religions with Christianity. Columbus had declared that the Indians 'would readily become Christians as they have a good understanding'.

Cortes and the Aztecs

The Aztecs were the ruling people of Mexico. When they first saw Spanish ships sailing past their coast, they described them as 'towers or small mountains floating on the waves of the sea.' In 1518 Hernando Cortes led an expedition into the interior of their country. He had five hundred soldiers, but they were armed only with two small cannons and forty-eight muskets, and had sixteen horses. However, he was immediately supported by the native tribes that lived under the rule of the Aztecs, who demanded from them regular human sacrifices to their gods, sometimes as many as 20,000 in a single year.

33

The Aztecs themselves could not decide whether the Spaniards were friendly messengers sent by the gods or a hostile army. When Cortes led his men into the Aztec capital, Tenochtitlan, its citizens sought to appease the strangers by sprinkling their food with the fresh, bright blood of human sacrifices. To their confusion the Spaniards 'closed their eyes and shook their heads in abhorrence.' The Aztec chieftain,

Cortes is received by Montezuma, King of the Aztecs. Montezuma was later killed during the desperate struggle between the Aztecs and the Conquistadores.

Tenochtitlan, the ancient capital of the Aztec Empire, on the present site of Mexico City.

Montezuma, welcomed Cortes, who wrote, 'When I approached to speak to Montezuma, I took off a collar of pearls and glass diamonds that I wore and put it on his neck, and after we had gone through some of the streets, one of his servants came with two collars, wrapped in a cloth, which were made of coloured shells ... When he received them, he turned towards me and put them on my neck.'

The city, which was larger than any town in Europe at that time, astonished the Spaniards. It was built on islands in a lake, and was surrounded by mountains and volcanoes. They had expected to find naked savages living in primitive conditions, but thay saw 'things that had never been heard of or seen before, not even dreamed about', including fine palaces filled with gold, silver and jade.

The Spaniards were alarmed by a great Aztec dance in honour of their gods. They began to massacre the townspeople, who stoned Montezuma to death when he tried to calm them. The Spaniards fought their way out of the city with heavy losses, but Cortes recaptured it after bitter fighting. Most of the buildings were destroyed during the battle, but Cortes at once began to rebuild them. 'Your Sacred Majesty may believe,' he wrote to the King of Spain in 1524, 'that within five years this city will be the most nobly populated city which exists in the world and will have the finest buildings.'

Pizarro and the Incas

The Incas possessed a great empire covering Peru, much of Ecuador, Chile, Bolivia and north-western Argentina. They were an even more cultured and wealthy race than the Aztecs. Their conqueror was Francisco Pizarro, a brave but cruel man. He told his 180 men, as they set out in 1531, 'Friends and comrades! On that side are toil, hunger, nakedness, the drenching storm, desertion and death; on this side ease and pleasure. There lies Peru with its riches; here Panama and its poverty. Choose, each man, what best becomes a brave Spaniard. For my part, I go to the south.'

Atahuallpa, the last Inca Emperor of Peru, who was put to death by the Spanish under Pizarro.

The great Temple of the Sun in Cuzco was desecrated by Pizarro and his soldiers.

The Incas were at this time divided by a civil war, which assisted the Spaniards in their conquest. Their ruler, Atahuallpa, hoped to win the support of the Spaniards against his enemies, but instead they responded with violence and treachery. Pizarro seized Atahuallpa, who offered as his ransom golden treasures that filled a room twenty-two feet long, seventeen feet wide and nine feet high, but this was not enough. Atahuallpa was put to death, and, by 1533, the Spaniards had conquered the country with little resistance. The great Temple of the Sun in Cuzco, the capital city, was desecrated and stripped of its gold. In 1541 Pizarro was murdered in his own house during a quarrel among the Spaniards.

Cortes was amply rewarded for having defeated the Aztecs and for having conquered Mexico.

The Spanish Empire

By 1540 all of South America, with the exception of Brazil, was under Spanish rule. The conquistadores were rewarded by the Spanish monarchy with large estates. Cortes' estate contained some 100,000 Indians and spread over 25,000 square miles. These estates grew valuable crops such as cotton, tobacco and sugar, but they also yielded precious metals which brought great wealth. After the accumulated treasures of the Aztecs and Incas were plundered, gold and silver mines steadily increased their production.

At first the native peoples laboured in the plantations and mines, but soon whole villages were wiped out by malaria, smallpox, measles

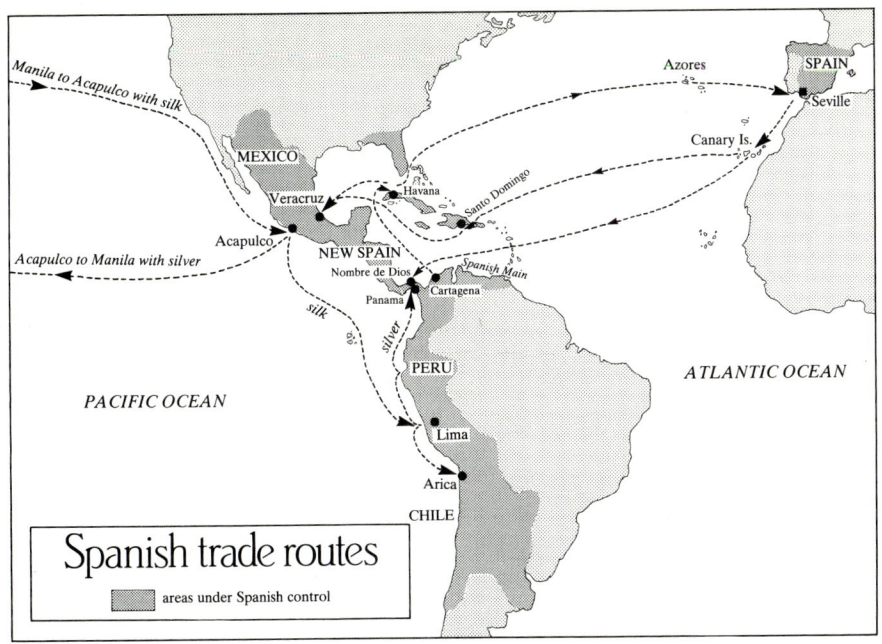

Manila to Acapulco with silk

SPAIN
Azores
Seville
Canary Is.

MEXICO
Veracruz
Havana
Santo Domingo

Acapulco
Acapulco to Manila with silver
NEW SPAIN
Nombre de Dios
Spanish Main
Panama
Cartagena
silk
silver

PERU
ATLANTIC OCEAN

PACIFIC OCEAN

Lima

Arica
CHILE

Spanish trade routes

areas under Spanish control

and other diseases that had been brought from Europe by the Spaniards. To replace them, black slaves were brought out from West Africa. By 1600 there were already about 40,000 of them in South America.

The Spanish monarchy, however, was not prepared to let the conquistadores exercise political power. Instead it was represented by the Viceroy, who governed through royal officials established in fortified towns. As with Portugal, the monarchy also controlled all imperial trade. Queen Isabella had asserted that since the empire had been discovered and conquered 'at the expense of my kingdom and settled with the nationals of my kingdom, it is right that all their trade and traffic should belong to my kingdom of Spain and be conducted from it; and that everything brought from the Indies should go to it and be for it.' Consequently South American exports were taken to Spain alone by means of two great 'treasure fleets' a year, which sailed under royal supervision.

'We Spaniards,' said Cortes, 'suffer from a disease that only gold can cure.' The 'disease' was the expense of Spain's determination to maintain her supremacy in Europe during the sixteenth century. She needed the precious metals of the New World to finance the building of ships, the construction of fortresses, and also to pay soldiers and fight wars. She could continue to do this as long as the stream of gold and silver was produced by her American mines.

6 THE CHALLENGE OF THE DUTCH

The wealth and power gained from their overseas possessions by Portugal and Spain inevitably attracted the envy of other European nations. The first to challenge the supremacy of these two empires were the Dutch, who remained their most determined opponents until the middle of the seventeenth century.

The Dutch expansion

The period of Dutch greatness began with their independence, which they achieved in 1572, when they rose in revolt against their Spanish rulers. The revolt was led by a group of seamen, who adopted the name of the 'Sea Beggars'. They raided Spanish ships and, after Spain had conquered Portugal in 1581, they set out to challenge the

Dutch 'Sea Beggars' set out to challenge the Portuguese claim to a monopoly of trade in the East Indies.

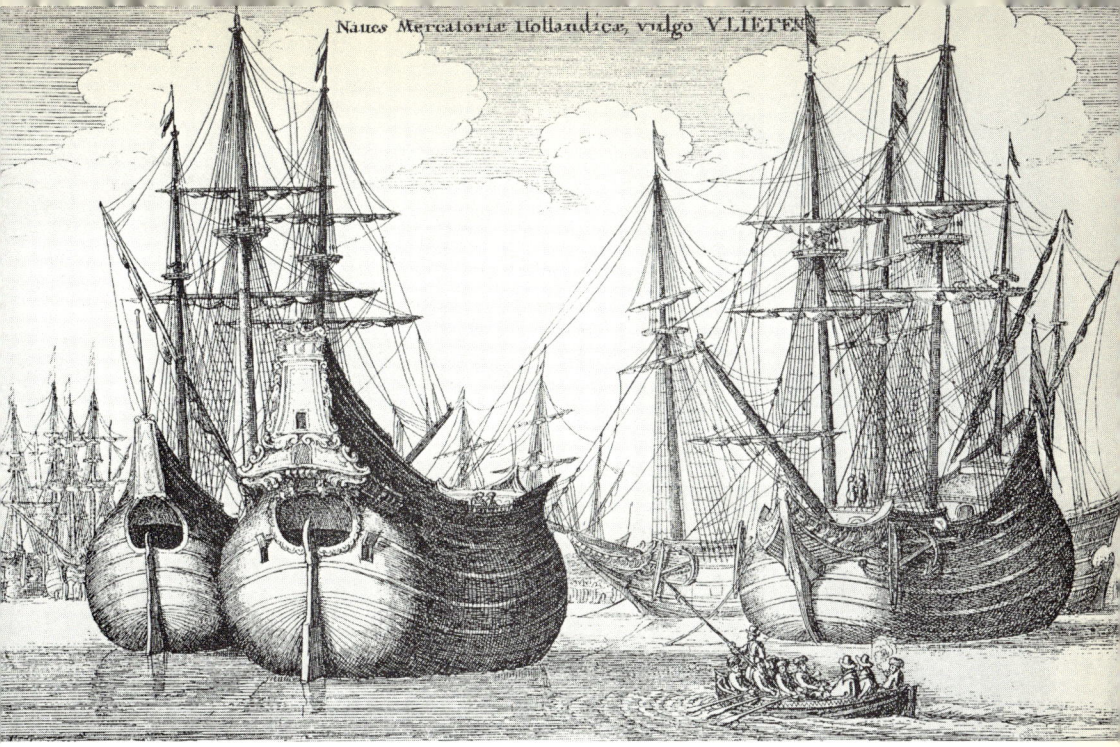

Trading vessels of the Dutch East India Company.

Portuguese claim to a monopoly of trade in the East Indies. When peace was made with Spain in 1609, some Dutch merchants protested that their profits could not 'be accomplished by the trifling trade with the Indies or the tardy cultivation of uninhabited regions, but in reality by acts of hostility against the ships and property of the King of Spain and his subjects.'

In addition to the determination of their seamen, there were other reasons for Dutch greatness. The Dutch trading companies were more successful than those of other nations. They were never short of money, as Amsterdam became the great financial centre of Europe after the establishment of the national bank and stock exchange there in 1609. Also Holland was the most tolerant country in Europe. It became Protestant after gaining independence, but other religions were allowed complete freedom. Foreigners were allowed to settle there and bring their money and talents. Andrew Marvell, the English poet, said that in Amsterdam were to be found 'Turk, Christian, Pagan, Jew', working and trading together.

The East India Company
From 1595 the Dutch began to send fleets to India, and by 1602 sixty-five of their ships had made the round voyage, bringing back silk and spices. The profits were immense, but so also were the risks, as the Portuguese regarded Dutch intrusion as an act of war. Few individual

The Dutch East India Company was forced to engage in naval and military action in their quest to establish a trading empire.

traders could afford to engage in the venture, but the formation of the Dutch East India Company, in 1602, solved this problem.

The Company was able to send out more ships, to establish trading-posts and forts, and to make treaties with local rulers. Gradually the Dutch supplanted the Portuguese, who were weakened by inefficiency and corruption to such a degree that they could not resist the more capable Dutch merchants. Holland, like Portugal, had too small a population to send large numbers of emigrants overseas. The Dutch also wanted to establish an empire of trade and did not seek territory for settlement.

The directors of the Company ordered their agents 'to keep in view the necessity of peaceful trade in Asia from which is derived the smoke in the kitchens here at home.' However, the Company had to engage in naval and military action and to occupy territory. Early in the seventeenth century, it took over Java, Sumatra, Borneo and the Spice Islands. From 1616 it established posts on the Indian mainland, and by 1658 had accomplished the conquest of Ceylon. Jan Pieterszoon Coen, who was appointed Governor-General of the Dutch Empire in 1618, adopted a more aggressive policy. He sent ships to trade with China and Japan. He also made Batavia, on the northern coast of Java, the Dutch capital, and in 1629 died defending it against the Javanese. His policy was continued by his successor, Anthony Van Diemen, under whom Formosa was occupied in 1642. In the same year, the

Abel Tasman discovered Tasmania, New Zealand, and the Tonga and Fiji Islands.

Dutch navigator, Abel Tasman, discovered Australia and New Zealand, but the Company did not possess sufficient resources to exploit these territories.

The Dutch soon began to engage in slave trading on the Guinea coast as they journeyed to the East, and in 1648 one of their ships was wrecked near the Cape of Good Hope. When the crew were rescued, nearly a year later, the description of the favourable climate and rich soil of the region encouraged the Company to establish a station there, which supplied food to its passing ships. Vines were planted in this area in 1655, and the Cape became one of the few Dutch possessions settled by colonists.

Piet Hien, who led the Dutch capture of the entire Spanish Treasure Fleet as it lay in a Cuban Harbour.

The West India Company

In the West, Dutch seamen raided settlements, captured ships, and smuggled slaves into the Spanish and Portuguese colonies in America. In 1621, when war was renewed against Spain, the West India Company was formed and provided with a fleet of twenty warships by the Dutch government. The Company established trading-posts in Guyana and also on the River Amazon. In 1628 Pernambuco was captured and held for twenty-six years, threatening the Portuguese occupation of Brazil. More profitable, however, were the attacks on Spanish shipping. The most notable of these raids occurred in 1628, when Piet Hien

Henry Hudson, the English navigator who explored the Hudson River in 1609 and Hudson Bay in 1610, where his crew mutinied and cast him adrift to die.

captured an entire treasure fleet, as it lay in a Cuban harbour. This enabled the Company to pay a dividend of fifty per cent on its capital that year.

In 1609 the East India Company had hired Henry Hudson, an English navigator, to find a north-west passage to the East. He managed to get no further than a 'great stream,' now called the Hudson River, near present-day Albany in North America. The East India Company took no interest in his discovery, but in 1624 the West India Company established the colony of New Netherlands at Albany, and in the following year the settlement of New Amsterdam was set up on Manhattan Island. Any member of the Company who could persuade fifty families to settle there was offered sixteen miles of land along the river 'and would forever possess all the lands . . . fruits . . . and lower jurisdictions' of the site.

The settlement of New Amsterdam, present day New York, in 1673.

For many Dutchmen, however, the attractions of farming and lumbering were far outweighed by the riches to be obtained by trading in the East Indies. The colony never had more than 7,000 inhabitants, and it proved difficult to find anyone of sufficient calibre to give leadership. A settler there complained, 'I told the Secretary that I was surprised the West India Company sent such fools to the country who knew nothing but how to drink themselves drunk. In the East Indies they would not be allowed to serve even as assistants.' Such Governors were incapable of making the colony a strong, self-governing community. Peter Stuyvesant, the last and most capable of these Governors, declared, 'We derive our authority from God and the Company, not from a few ignorant subjects.' The Dutch attempt at establishing an American empire never seemed likely to succeed.

7 THE ENGLISH IN NORTH AMERICA

When Portugal and Spain were founding overseas empires and discovering new trade routes, England was a weak and isolated country. She was still recovering from the Wars of the Roses, which had been a long-drawn-out civil war, and she was only beginning to build up a merchant fleet of her own during the reign of the first Tudor ruler, Henry VII (1485–1509).

Tudor seamen
It is not surprising, therefore, that the first Tudor seaman of note was of Italian origin, John Cabot. He sailed from Bristol, in 1497, to search for the 'island of Brazil to the westward of Ireland.' When he returned he claimed that he had reached Asia and had discovered the 'mainland of the country of the Great Khan', where Marco Polo had been. He had, in fact, discovered North America and had probably seen the island of Newfoundland. This was to prove an important part of the world for England once she had developed into a stronger nation.

A map showing Cabot's two voyages across the Atlantic to the north coast of America.

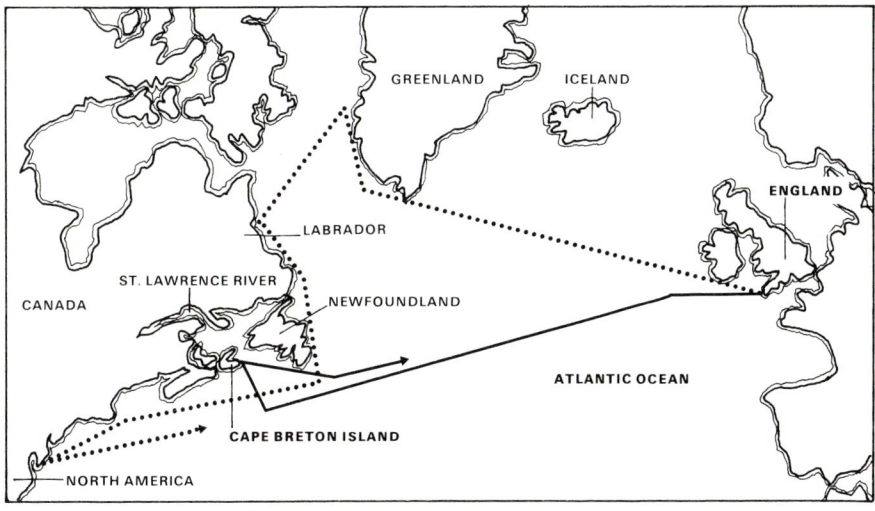

ÆTATIS SVÆ LVIII
Anno Dmi 1591

Sir John Hawkins, the English naval commander and slave trader, who was commander of a squadron in the fleet that defeated the Spanish Armada in 1588.

Not until Elizabeth I's reign (1558–1603) did English seamen sail across the Atlantic. In 1562, John Hawkins sailed with three ships to West Africa, where he 'got into his possession, partly by the sword and partly by other means, to the number of 300 negroes at the least.' These he took to Spanish America and sold as slaves so 'that he did not only load his own ships with hides, ginger, sugars and some quantity of pearls, but he freighted two other hulks with hides.' He made another equally profitable voyage two years later, but his third voyage in 1567, ended disastrously when he was attacked off Mexico by the Spaniards and lost two of his four ships. Spain was determined to exclude English ships from South American waters, and the two nations became enemies. They were further divided for religious reasons because England was now a Protestant country and Spain was one of the foremost Roman Catholic countries.

The most famous Tudor seamen was Francis Drake, who was the first Englishman to sail round the world. In 1577 he surprised a Spanish

Francis Drake raises a toast to Queen Elizabeth I.

Sir Walter Ralegh introduced tobacco and potatoes into England.

treasure fleet in the Pacific and took so much from it that his ship, the *Golden Hind*, was actually ballasted with silver bars. He sailed northwards as far as California, which he claimed for the Queen as 'New Albion', and then struck out across the Pacific, rounded the Cape of Good Hope and arrived back in England after a voyage of nearly three years. The Queen knighted him, saying, 'Your actions have done you more honour than the title which I have conferred.'

Jamestown and Bermuda

English seamen made several unsuccessful attempts to establish colonies in North America. In 1585, Sir Walter Ralegh sent settlers to a colony which he called Virginia in honour of Queen Elizabeth, but within a year they had grown dissatisfied with conditions there and returned home. He sent another group to what is now North Carolina in 1587, but because England was involved in a war with Spain in the following year, it was not until 1591 that supply ships reached the colony. The relief vessels 'sounded with a trumpet a call and afterwards many familiar English tunes of songs and called to them friendly, but had no answer.' The settlers had mysteriously disappeared, leaving behind only deserted huts.

In 1607 the Virginia Company was founded in London, and it established the first permanent colony in an area that is now called Chesapeake Bay. It was named Jamestown, in honour of James I, and initially populated by about a hundred people. They faced terrible hardships. By 1611, out of a total of nine hundred settlers that had arrived at the colony only 150 were still alive. Most had died by disease, or been killed during attacks from the native Indians. Indeed, this colony too might have failed, had it not been for a young man, John Smith, who made the settlers grow crops and erect defences against the Indians. On one occasion the Indians threatened to kill him, but 'Pocahontas, the chief's dearest daughter, when no entreaty could prevail, got his head in her arms and laid her own upon his to save him from death.' Later Pocahontas married another settler, John Rolfe, and went with him to England.

Rolfe was the first settler to discover that tobacco could be grown in Virginia, and in 1619 the first African slaves were brought there to work on the plantations. Tobacco made Virginia prosperous and other southern colonies that were later established by England in America were to profit similarly. Every year tobacco was packed in huge hogsheads, and shipped to Bristol and Liverpool. Smoking became increasingly common in England, though James I condemned it as 'harmful to the brain, dangerous to the lungs'.

In 1609 an English ship was wrecked on the uninhabited island of Bermuda in the West Indies. Some of the crew built another ship and

In 1620, because of the lack of females in Virginia, ninety young women were induced to seek husbands and their fortunes in the colony.

sailed away to Virginia, but others preferred to stay there and were joined by settlers from England. John Smith visited the island and praised its climate: 'No cold there is beyond an English April, nor heat much greater than an ordinary July in France, so that frost and snow is never seen here, nor stinking and infectious mists very seldom.' Tobacco became Bermuda's main crop, and black slaves were sent there. The discovery of Bermuda inspired Shakespeare to write *The Tempest*, his play about a shipwreck and a strange, wonderful island.

African slaves were transported to Virginia to work on the tobacco plantations.

The Pilgrim Fathers

In 1620 a group of about a hundred people, who have become collectively known as the Pilgrim Fathers, sailed from Plymouth in the *Mayflower*. They were Puritans, who were not allowed to worship in England as they wished and intended to seek a new life in Virginia. However their ship was blown far north, to Cape Cod Bay, where they established a settlement which they called Plymouth. They arrived in the winter when it was bitterly cold, and at first they found hardly any food beyond oysters and clams. Before the end of the year, half of the settlers had died from the effects of cold, exposure and starvation. However, they learnt how to plant corn as the Indians did, by putting fish for manure in the drill-holes. They built log huts and a fort armed with six cannon. When the *Mayflower* returned to England in the early summer, not one settler chose to go with her.

The Pilgrim Fathers were Puritans who were not allowed to worship in England as they wished, so they sought a new life in America.

'The Mayflower' on which the Pilgrim Fathers sailed.

The *Mayflower* took a letter from one of the Pilgrim Fathers, urging English people to join them; 'The country wanteth only industrious men to employ. For it would grieve your hearts if you had seen so many miles, by good rivers, uninhabited.' Settlers did come from England, many of them Puritans. Other townships were formed in the region of Plymouth, and these united to form the colony of Massachusetts. After this other colonies were founded—Rhode Island and Connecticut southwards, Maine and New Hampshire northwards. Most of the colonists were small farmers who owned their land and worked on it themselves. Others were engaged in lumbering, fur-trading and fishing. Though the climate was more extreme and the countryside more forested, these five colonies had so much in common with the mother country that together they were called New England.

8 THE FRENCH IN CANADA

Throughout the sixteenth century, French deep-sea fishermen sailed to the fishing-grounds off Newfoundland and the American coast. Sometimes they went ashore and found that, in exchange for tools and pottery, the Indians gave them furs which fetched a good price in Europe. French interest was aroused in this part of the world.

Jacques Cartier, the French navigator and explorer in Canada, who discovered the St Lawrence River.

The French settlement of Quebec founded in 1608 by Samuel de Champlain.

The fortunes of the fur trade

In 1535 a French explorer, Jacques Cartier, defied the constant dangers presented by drifting ice, and boldly sailed up the River St. Lawrence. He ventured as far as the future site of Montreal and saw something of the vast interior of Canada. The French government wanted to establish a permanent settlement there, in order to develop the fisheries and fur trade. But France was torn by a civil war, the Wars of Religion, and it was not until 1604 that a royal expedition established the

Early settlers traded for fur with the Indians.

settlement of Acadia. Later the English claimed this territory, which was named Nova Scotia by James I, and attacked it several times, but the French were able to retain it.

More important, however, was the work of Samuel de Champlain, who was the first European to use Indian canoes for travelling along the inland waterways. He observed a place on the St. Lawrence where the steep cliffs from the bank made a natural fortress commanding the narrow bend of the river and, in 1608, he founded a settlement called Quebec there. This was a trading-post, rather than a colony. Its inhabitants traded with the Indians and obtained furs, which they sent down the river to be shipped to France.

Other settlements were established further up the river. The most important of these was Montreal, which was founded in 1642. But the colony of New France remained small and weak. In 1666 its population numbered less than 4,000, while that of the English colonies had reached over 50,000. Most of the French in Canada were still fur-traders and in the late 1640s they suffered from an outbreak of warfare between the Indian tribes. Champlain had made an alliance with the Hurons, but the English supported their enemies, the Iroquois. The Iroquois ambushed the fur-laden canoes on the waterways so successfully that by 1652 they had brought the trade almost to a standstill. An observer said, 'The beavers are left in peace and in the place of their repose. The Huron fleets no longer come down to trade. ... In the Quebec warehouse there is nothing but poverty.'

The Company of the West
The situation changed, however, in 1661, when Louis XIV began to govern France. He was determined to assert the power of France over other nations. The treasure of the New World had produced the same harmful effects upon Spain as it had upon Portugal, and now the Spanish gold and silver mines were becoming exhausted. Spain's supremacy in Europe was weakening, and Louis XIV fought a series of wars during the remaining years of the century that sought to challenge it.

During the first part of Louis' reign his Controller-General of Finance, Jean Colbert, attempted to make France a wealthier country. This policy included encouraging the expansion of French trade and power overseas. He wished to develop a French colonial empire in North America. In 1664 he established the Company of the West, which directed all French settlements in North America, the West Indies and Africa. It received a grant from Louis XIV of the whole of North America, extending from the Hudson Bay to Florida, despite the existence of the English and Dutch colonies on the Atlantic coast. Furthermore the administration of New France was placed under the French Crown.

Louis XIV, King of France, who was known as 'The Sun King'.

With the establishment of the Company of the West, French rule in Canada was strengthened and the Iroquois were subdued.

In the following years, French rule in Canada was greatly strengthened. The Iroquois were subdued, and fur-trading revived. Emigration was encouraged and the white population increased rapidly, so that by 1679 it numbered 10,000. Efforts were made to supplement fur-trading by agricultural production along the great river from Quebec to Montreal, and in seven years the area of cultivated land in this region was doubled. Despite this expansion, Louis XIV regarded the wars in Europe as his most important area of concern and consequently he began to divert resources in that direction. In 1666 Colbert had to tell officials in Canada that Louis thought it unwise to weaken France by sending further manpower across the Atlantic, and in 1672 he had to report that there was no government money to spend on Canada. So the hope of making Canada a great French colony declined. The fur-trade remained the chief occupation of the white population, which as a result naturally became dispersed, leaving few important centres of settlement. French colonization in North America did not progress beyond a few scattered forts, a number of adventurous fur-traders and an occasional, small village of white farmers and merchants.

The expansion of French rule

Nevertheless, French rule in North America was extended. The wealth that the Spaniards had found in South America meant that they took little interest in the West Indies. But these islands grew in importance when sugar-cane crops were introduced there from Brazil in the 1650s. Sugar replaced tobacco as their main crop. The French followed the English example and introduced it in their islands of Martinique, Guadeloupe and Dominica. Frenchmen were more willing to go to the West Indies than to Canada. About a thousand immigrants a year came from France, and French trading-posts were established in West Africa to organize the transportation of slaves to work in the sugar-plantations. By 1685 the population of the islands was estimated to be 52,000, of which about two-thirds were black slaves. Between 1662 and 1683, the number of French ships engaged in the sugar trade rose from four to over two hundred. In France, sugar refineries were established and the ports of La Rochelle, Nantes and Bordeaux were developed in order to deal with this increase in traffic. In addition the West Indies provided a good market for French exports such as wine, brandy, olive oil and cloth.

African slaves were transported to the West Indies to work in the sugar plantations.

Robert Cavalier, Sieur de la Salle, who founded Louisiana.

There was also an expansion of French territory on the American mainland, as both the fur-traders and Jesuit missionaries penetrated beyond the Great Lakes to the Mississippi River. This forward movement reached its limits in 1682, when Robert Cavalier, Sieur de la Salle, arrived at the Gulf of Mexico and founded Louisiana. Great hopes were placed upon this colony, which was named after Louis XIV, but the success of the West Indies was not to be repeated there. The sub-tropical climate, with its heavy rainfall, did not suit French tastes. When the town of New Orleans was founded a hundred miles up the winding Mississippi river, its inhabitants complained, 'If the river doesn't get you from the front, Lake Pontchartrain waters you from the back.' Attempts to grow tobacco, sugar and cotton failed through the inexperience of the settlers and a lack of adequate manpower.

In addition to these problems, the wide extension of French territory in North America, without a population sufficient to settle it properly, was very dangerous. English resentment and fears were aroused. One Governor of Montreal thought that there was bound to be a clash between French and English ambitions on the continent. He wrote, 'It would be difficult for our colony or theirs to subsist other than through the destruction of the other.'

The English founded several factories in the East Indian Islands.

9 OVERSEAS RIVALRY

The judgement of the Governor of Montreal was shown to be only too true. In both East and West, the new colonial powers—Holland, England and France—found themselves in competition with each other and frequently clashed in their determination to expand overseas trade and settlements.

The East Indies

In 1599, a meeting of some London merchants 'in my Lord Mayor's parlour to consider the unchristian price of pepper', led to the formation of the East India Company to engage in the eastern trade. Soon its ships were returning with profitable cargoes. The Portuguese did not like this competition in their market and told Indian rulers not to trade with England. One of them 'most vilely abused His Majesty [King James I], terming him King of the Fishermen and of an island of no importance.' But, in 1612 a fleet of armed ships sent out by the Company defeated the Portuguese in a decisive battle at Swally Roads, which hastened the decline of Portuguese involvement in the eastern trade. The Company now founded several 'factories' (or trading-posts) on the East Indian Islands, but this increased the determination of the Dutch to drive them out of the region. In 1623 they massacred all the English merchants at a factory on Amboyna, one of the Spice Islands.

The English were now forced to go to the mainland of India. By the end of the century they had established three factories there—Bombay on the west coast, Madras on the east coast, and Calcutta in Bengal. However, the Company did not want to become involved in Indian affairs. It tried to follow the advice of one of its officials—'If you will profit, seek it at sea and in quiet trade for ... it is an error to affect garrisons and land wars in India.'

Some years later France also sought to trade in India. Colbert founded an East Indies Company in 1664. The Company built several factories, the most important being at Pondicherry, south of Madras. Initially the French and English companies believed that there was room for them both to trade in India, and there was no conflict between them during the seventeenth century.

The Anglo-French conflict in America

The situation in America, however, was very different, and there was constant conflict between the two nations. In the 1660s two Frenchmen, the Sieur de Groseillers and Pierre Radisson, travelled north to Hudson Bay and tried in vain to persuade the Company of the West to extend its fur-trading activities there. Finally, they went to England, where they were nicknamed 'Gooseberry and Radishes', and the result was the formation, in 1670 of the Hudson's Bay Company. Each year it sent ships to the Bay loaded with weapons, trinkets and utensils to barter with the Indians in exchange for high-quality furs. The French realized the error of their earlier decision and began to send their traders to the territory in order to compete with the English.

The rivalry between the English and French colonists was made even more bitter and violent in America. The English had now established twelve colonies on the continent. They had captured New Amsterdam from the Dutch in 1664 and renamed it New York in honour of James, Duke of York, who was later to become King James II. Furthermore they had expanded their territory to the south as far as the Carolinas.

The Hudson Bay Company's fleet, 'Prince Rupert', 'King George', and 'Sea Horse'.

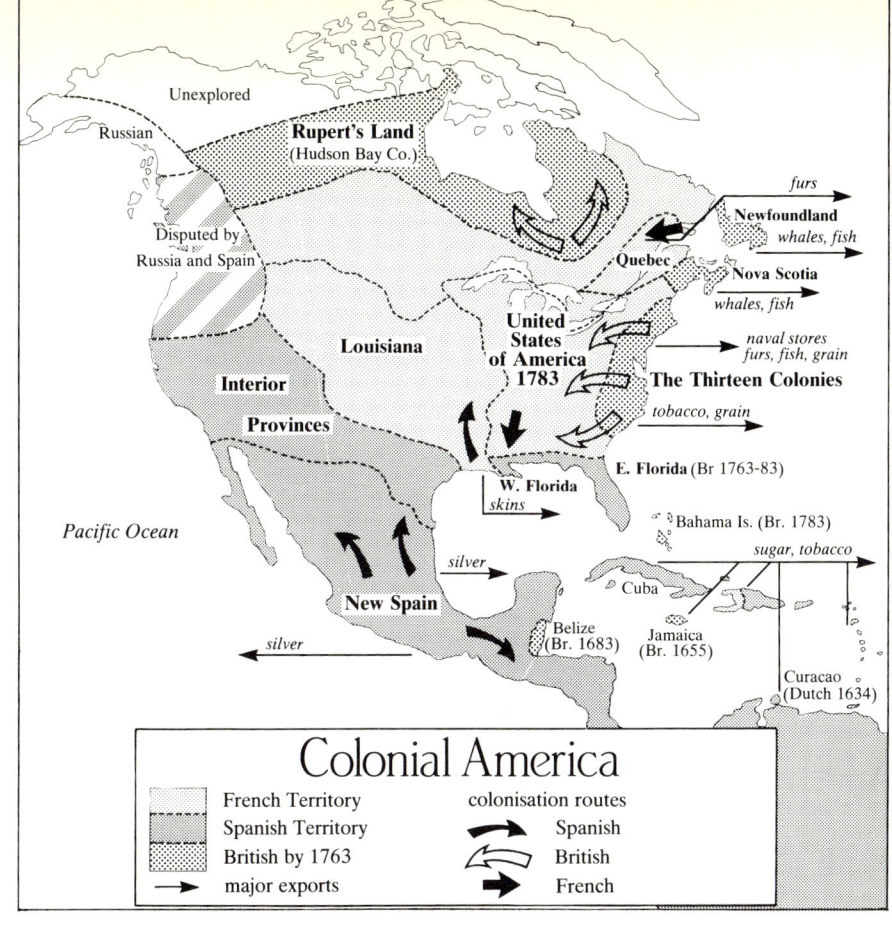

Colonial America

▧	French Territory	colonisation routes
▨	Spanish Territory	➤ Spanish
▦	British by 1763	⇨ British
→	major exports	➡ French

The French felt themselves threatened and la Salle warned his country-men that the English could 'complete the ruin of New France, which they had already hemmed in by their establishments in Virginia, Penn-sylvania, New England and Hudson Bay.'

For the French it was extremely important to maintain their crescent-shaped line of forts that skirted the St. Lawrence, the Great Lakes, the Ohio and the Mississippi. These could be used to prevent the Eng-lish colonists moving into the interior, while enabling them to advance past Lake Champlain and the Hudson valley. If they did this they would drive a wedge through the English colonies and make it possible to conquer them. The English colonists, for their part, tried to capture the French forts and so penetrate their defences.

No fixed frontier separated the settlements of the two nations, but the outposts of each side crept steadily closer to each other. Both sides made alliances with Indian tribes, but because the French were nearly all fur-traders and were on good terms with the Indians, they were in a much better position to do this. The English were mostly farmers and settlers who had aroused resentment among the Indians by taking land from them. With the exception of the Iroquois, all the

The English farmers and settlers aroused resentment among the Indians.

Indian tribes were French allies. Mixed parties of French and Indians burned English villages and killed the inhabitants. The best-known of these incidents occurred on a winter's night at Deerfield in Massachusetts, when a horde of Indians and French Canadians climbed over the snow-drifted palisade of the settlement, attacked the inhabitants in their beds and set the wooden houses alight. They left the village a smoking ruin, with half its 290 inhabitants either killed or dragged off to the north. A commemorative stone that stands in the town's churchyard to-day still bears the inscription, 'The Tomb of 48 Men, Women and Children, Victims of the French and Indian Raid on Deerfield, February 29, 1704.'

The English retaliated as best they could. The waterways of the Penobscot, Kennebec, Connecticut and Hudson were well-suited for attacks directed towards the St. Lawrence. Savage incidents took place along these rivers and upon the shores of Lake Champlain. The overall advantage in the struggle lay with the English, as they had both a larger population and possessed greater military strength. France was handicapped by the need to protect her frontiers in Europe and could not send effective help overseas. In a year, for example, when Louis XIV had had to maintain 200,000 soldiers and a large fleet in Europe, he could send Canada only sixty girls as wives for settlers.

The War of the Spanish Succession (1702–13)
The colonial border-fighting between the French and English was almost continuous, but it worsened when the two countries were formally at war. Such a conflict was the War of the Spanish Succession, in which England fought against France and Spain. French weaknesses became very clear at this time, as she had to take part in heavy fighting in Europe and also divide her fleet between the Mediterranean and the Atlantic. England, on the other hand had no need for big armies and could concentrate most of her fleet in the Atlantic and American waters. Consequently she could send supplies and reinforcements to her colonies and also cut off those for the French colonies.

The Treaty of Utrecht, which ended the war in 1713, brought England overseas gains. She had seized Gibraltar and Minorca, which she kept as Mediterranean naval bases, as well as Acadia, which now became Nova Scotia. France formally recognized the English right to Newfoundland and Hudson Bay, and also acknowledged her alliance with the Iroquois, which brought the boundary of the colony of New York to Lake Ontario. Furthermore the war revealed England's worldwide naval supremacy. The American historian, A. T. Mahan, has said, 'Before that war England was one of the sea-powers. After it, she was *the* sea-power, without any second.' During the next two centuries, this was to gain her the greatest overseas empire in the world.

The colonial border warfare between the French and the English was almost continuous.

GLOSSARY

Almanac   An annual book of tables, containing a calendar and information and calculations about the stars and other heavenly bodies.

Bazaar   An eastern market-place, consisting of many shops or stalls-where different kinds of merchandise are offered for sale by traders.

Caravel   A small, light, partly-decked ship, which had a square sail on the foremast and lateen sails on the other two or three masts.

Carrack   A larger, fully-decked ship, which had square sails on the fore and main masts, a lateen on the after mast and a sail slung under the bowsprit.

Cartographer   A man who makes maps or charts.

Compass   An instrument, consisting of a magnetized needle turning on a pivot and pointing northwards, which is used in the guidance of a ship's course at sea.

Hogshead   A large cask, usually holding about fifty gallons.

Quadrant   An instrument, used for measuring angles in astronomy and navigation, which has a wooden frame with a movable wooden arm for sighting the objects to be measured.

Renaissance   The great revival of art, literature and science, under the influence of the works of ancient Greek and Latin thinkers, which took place in Europe between the fourteenth and sixteenth centuries.

Scurvy   A disease caused by lack of vitamin C found in fresh food, especially fruit and vegetables.

Yard   A long beam on a mast for spreading square sails.

DATE CHART

1415 Portuguese capture of Ceuta

1419–60 Henry the Navigator at Sagres

1420 Portuguese discovery of Madeira

1431 Portuguese discovery of the Azores

1441 Portuguese discovery of the Cape Verde Islands

1487 Bartholomew Diaz reached the Cape of Good Hope

1492 Christopher Columbus crossed the Atlantic

1497–99 Vasco da Gama's voyage to Calicut

1497 John Cabot discovered Newfoundland

1500 Portuguese annexation of Brazil

1511 Portuguese capture of Goa

1513 Balboa's discovery of the Pacific

1518–21 Cortes's conquest of the Aztecs

1519 Magellan reached the Pacific

1531–33 Pizarro's conquest of the Incas

1535 Cartier sailed up the St. Lawrence

1562–67 Voyages of John Hawkins

1577–80 Drake's voyage around the World

1588 English defeat of the Spanish Armada

1599 English East India Company founded

1602 Dutch East India Company founded

1607 Jamestown founded

1608 Quebec founded

1609 Bermuda occupied

1612 Battle of Swally Roads (India)

1620	Pilgrim Fathers sailed to New England	1664	French Company of the West founded
1623	Amboyna Massacre		French East Indies Company founded
1624	New Netherlands founded		English captured New Amsterdam (New York)
1639	English factory at Madras		
1642	Dutch conquest of Formosa	1670	Hudson's Bay Company founded
	Tasman's discovery of Australia and New Zealand	1674	French factory at Pondicherry founded
	Montreal founded	1682	Louisiana founded by la Salle
1651	Dutch settlement of the Cape of Good Hope	1684	English factory at Calcutta founded
1658	Dutch conquest of Ceylon	1702–13	War of the Spanish Succession
1661	English factory at Bombay founded	1713	Treaty of Utrecht

PICTURE ACKNOWLEDGEMENTS

The illustrations in this book were supplied by: Mary Evans 5, 12, 15, 30, 32, 35, 37, 38, 41, 44, 57, 61, 62; Michael Holford *front cover*, 4: Mansell Collection 6, 10, 11, 14, 17, 19, 21, 24, 25, 26, 27, 28, 31, 33, 34, 42, 43, 45, 46, 48, 49, 52, 54, 55, 56, 58, 60, 63, 64, 66, 70; Ann Ronan Picture Library 8, 36, 50, 53; Wayland Picture Library 40, 47, 68; Malcolm S. Walker 7, 16, 18, 20, 22, 29, 31, 39, 67.

FURTHER READING

Histories
Boxer, C. R. *The Portuguese Seaborne Empire, 1415–1825* Penguin
  Books, 1973
Boxer, C. R. *The Dutch Seaborne Empire* Penguin Books, 1973
Crone, G. R. *The Discovery of America* Hamish Hamilton, 1969
Diaz, Bernal (trans. J. M. Cohen) *The Conquest of New Spain*
  Harmondsworth, 1963
Elliott, J. H. *The Old World and the New, 1492–1650* Cambridge
  University Press, 1970
Gardner, Brian *The East India Company* Hart Davis, 1971
Hemming, John *The Conquest of the Incas* Macmillan, 1970
Innes, A. D. *The Maritime and Colonial Expansion of England under
  the Stuarts, 1603–1714* Low, 1931
Innes, Hammond *The Conquistadors* Collins, 1970
Kirkpatrick, F. A. *The Spanish Conquistadors* Black, 3rd. edn., 1963
Landstrøm, Bjørn *Columbus* Cambridge University Press, 1968
Nevins, Allan *A History of the American People from 1492* Oxford
  University Press, 2nd edn., 1970
Parker, G. *Spain and the Netherlands, 1559–1659* Fontana, 1979
Parry, J. H. *The Age of Reconnaissance* Weidenfeld & Nicolson, 1963
Parry, J. H. *The Spanish Seaborne Empire* Penguin Books, 1973
Saur, C. O. *The Early Spanish Main* University of California, 1966
Sterling, Thomas *Exploration of Africa* Cassell, 1964
Williamson, J. A. *The Age of Drake* Black, 5th. edn., 1965

Memoirs and Documents
Greenlee, W. B. (trans. & ed.) *The Voyage of Pedro Alvares Cabral
  to Brazil and India* Hakluyt Soc., 1938
Hollingworth, G. E. (ed.) *The World Encompassed* by Sir Francis
  Drake, Blackie, 1935
Jane, Cecil (trans.) *The Journal of Christopher Columbus* Blond, 1960
Keen, Benjamin (trans.) *The Life of the Admiral Christopher
  Columbus by his son*, Rutgers University Press, 1959
Lay, C. D. (ed.) *Portuguese Voyages, 1498–1663* Dent, 1947
Parry, J. H. (ed.) *The European Reconnaissance* [Documents]
  Macmillan, 1968
Ravenstein, E. G. (ed.) *The Journal of the First Voyage of Vasco da
  Gama, 1497–99* Hakluyt Soc., 1898

INDEX

d'Albuquerque, Alfonso 23
del Almeida, Francisco 23
America, discovery of 29
de Angleria, Peter Martyr 29
Asia 23–4, 29–30
Atahuallpa 37
Atlantic 5, 11, 26, 29
Aztecs 33

de Balboa, Vasco Nunez 30
Battle of Swally Roads 65

Cabot, John 47
Cabral, Pedro 22
Calicut 21–3, 65
Cape of Good Hope 19, 43, 51
Caravel 10, 13, 15, 16
Carrack 10
Cartier, Jacques 57
Cavalier, Robert 63
de Champlain, Samuel 59
China 7, 23, 42
Colombus, Christopher 7, 20,
    26–9, 32–3
Conquistadores 32–9
Cortes, Hernando 33–5, 38–9

Deerfield Massacre 69
Diaz, Bartholomew 18–19
Dutch
    East India Company 41–3
    in the East 23
    West India Company 44–6
Drake, Sir Francis 49

East Indies 30, 41, 65

Ferdinand, King of Spain 26
French
    Anglo–French conflict 66–9
    Company of the West 59–61
    Fur Trade 57–9
    in Canada 56–63

da Gama, Vasco 18, 21
Golden Hind 51

Hawkins, John 49
Henry, Prince, 'The Navigator'
    12–17
Hien, Piet 44
Hudson Bay Company 66–7
Hudson, Henry 46

Incas 33, 36–7
India 7, 22–3
Iroquois Indians 59, 67–9
Isabella, Queen of Spain 26, 39

James I, King of England 51
Jamestown 51
Japan 7, 23, 27, 42
John II, King of Portugal 18
John II, Pope 20
John, Prester 13

Kublai Khan 7, 47

Lateen sail 10
Lisbon 13, 21, 22, 23
Louis XIV, King of France 59,
  69

Magellan, Ferdinand 30
Manhattan Island 44
Manuel I, King of Portugal 21–2
Mayflower 54
Mexico 33, 49
Montezuma, King of the Aztecs
  35
Montreal 57, 61, 65

New Amsterdam 46, 66
New England 55, 67
Newfoundland 47, 56
Nicholas V, Pope 17
Nina 26

Pieterszoon Coen, Jan 42
Pilgrim Fathers 54–5
Pinta 26
Pizarro, Francisco 36–7
Pocahontas 51
Polo, Marco 7, 26, 47

Quadrant 9
Quebec 59, 61

Ralegh, Sir Walter 51
Rolfe, John 51

Sagres Bay 13, 16
St Lawrence River 57, 67
Santa Maria 26, 28
'Sea Beggars' 41
Shakespeare, William 5, 53
Slave Trade 13, 17, 23, 39, 51,
  53, 62
Smith, John 51
Spain 6, 10
  in South America 29–39

Tasman, Abel 43
Temple of the Sun 37
Tenochtitlan 34
Tierra del Fuego 30
Tobacco 51–3, 63
Tordesillas, Treaty of 20

Vespucci, Amerigo 29
Virginia Company 51

War of the Spanish Succession
  69
West Indies 20

Xavier, St Francis 23